EVOLUTION

From Start to Finish

In less than

fifty paragraphs

Everything you ever wanted to know about Evolution...

Christopher Malden

'Evolution' started as a subject for general attention with Erasmus Darwin – Charles' grandpa.

1. Darwin (Charles) was supposed to be a cleric. The family married into the Wedgewood clan; well to do, respected. So when Darwin tired of religion and decided not to following in his father's footsteps to pursue the divinity

degree at Cambridge there was not too much panic; his brother had been to Cambridge before him and was flying the family's academic and religious flag.

2. So Darwin indulged his passion for molluscs and natural history. He leapt at the chance of acting as ship's Naturalist when Captain Fitzroy was commissioned by the Navy to survey and chart the coasts of S. America in the 'Beagle'.

3. In his trips ashore – some of them lengthy – in S.America he penetrated with a few companions deep into the interior. He found massive fossil bones. These he sent back as curios to London by returning British traders.

4. Darwin became known as 'the fossil man' almost the status of a circus impresario – Monstrous fossils were hugely popular in the London 'Scene', fuelling debates about the origin

of these strange creatures, supposed by many to be drowned in the great Biblical flood.

5. Darwin met the 'savage' people of Tierra del Fuego, saw weird forms of life, started thinking.

6. Around Cape Horn the beagle headed north and eventually reached the Galapagos Islands.

7. Here, different kinds of similar birds inhabited the separate islands. Their significance dawned only later;

Darwin actually forgot to label or date the specimens he brought back. Luckily, the crew had recorded data under Fitzroy's direction.

8. A theory of evolution slowly evolved, too, in Darwin's mind: he had a kind of allotment garden, bred pigeons, talked to breeders, swapped stuff, became a respected naturalist.

9. Luckily he had married Emma Wedgewood. Their combined resources

enabled the purchase of a family home, Down House in Kent. Unluckily, knowledge of closely related individuals marrying was limited or ignored through reasons of their close family ties; they had a severely disabled child whose brain damage resulted in early death.

10. Additionally, Emma's and Darwin's religious faith was tested to extreme limits. Darwin became sort of agnostic.

Emma's faith survived, but under great strain.

11. Meanwhile, Darwin's notions of evolution by small, advantageous changes was being carefully, meticulously honed, recorded, revised. He loathed, and was terrified of, public appearances.

12. But his writings and theories generated strong scientific support. In equally vehement and energetic terms, condemnation arrived

from the establishment and the church.

13. So Darwin suffered continual ill health, taking the waters at various spas. Only when Alfred Russell-Wallace sent him a paper outlining an astonishingly close parallel theory of evolution by natural selection was Darwin goaded, cajoled, persuaded by friends to publish.

14. The Royal Society agreed to host a joint paper, and several decades after Darwin's first ideas were jotted down, he published 'On the Origin of Species'.

After Darwin

15. The scientific community adopted and, even in the light of current knowledge of DNA, accepts the theory

of evolution as written by Darwin and Wallace

16. 2006, Provence. The first work offering a new interpretation is published: *Dangerous Mind – On the Origin of Pseudo species.*

17. One of the problems of Darwin's theory is that it does not fully account for the massive gap between the behaviour of man (*Homo sapiens*) and other primates and mammals.

18. Neoteric Evolution (from the Greek 'looking anew', in one interpretation) enables us to abandon the long-accepted Victorian view that places man at the 'top' of some hypothetical evolutionary tree.

19. Darwin and science view evolution in an unquestioning way, as a system of 'advance'.

20. Neoteric Evolution holds that *stasis* is the key descriptor of

evolutionary outcomes, meaning species that remain *unchanged* over millions of years are true representations of 'evolution' and its mechanisms and hence thereby challenges conventional ideas of 'evolutionary success'.

21. The Neoteric view does its best to avoid the anthropomorphic warping of reality: the natural world, especially evolution, cannot be measured, determined, or subject to, the human

view; reality appears distorted because the human is "pre-loaded" with notions of sequence and order.

22. For example, recent discovery of macaque-like species perfectly preserved as a fossil in an extinct volcano's crater from 47 million years ago was hailed as a 'human ancestor', i.e. part of the primate lineage 'leading to' humans. In fact, the species is alive and well today, unchanged.

23. Among hundreds of thousands of other species Sharks, too, have remained virtually unchanged for 650 million years. They have not evolved. Evolution is not 'the driver' of life.

24. Humans are very recent types; literally 'upstarts', we emerged from a primate precursor *Pan troglodytes*, a chimpanzee, with whom we share around 98% of DNA.

25. By fashioning tools around 3.8 million years ago (dates vary), humans have come from nowhere to dominate the biosphere within an 'eye-blink' of evolutionary time.

26. Neoteric evolutionary theory prefers the term 'proliferation' of species, not 'evolution', which implies 'improvement over previous condition'. The new theory holds that 'proliferation' better describes the emergence

of divergent species, of further effectiveness in species exploiting a particular niche resulting from change in the niche itself, its complexity and variety, not any measure of species 'superiority'. Nevertheless, 'Evolution' is the term most widely accepted.

27. The Neoteric theory holds that mutation is the agent of diversification enabling a species *marginally* to better exploit an environmental niche.

Profound change may have arbitrary (chance) beginnings; changes in climate or landscape, rising sea levels or desertification may benefit a group already carrying a mutation in their genetic profile - a new island appearing on an avian migration route may cause a population boom.

28. The niche 'weeds out' individuals that don't possess qualities enabling them to best exploit, best 'fit', the

niche in a manner as efficient as others. Only survival of offspring enshrines 'beneficial' mutations. Weaker birds are 'culled' by natural events such as storms at sea.

29. Individuals pass on to their descendants DNA expressing for an advantageous trait; more buoyant feathers in a migrant bird. Or the species itself may change and sideline this comfortable 'fit', leading to alternative 'solutions';

breeding individuals might pass on an ability to withstand lower temperatures, making migration unnecessary.

30. But, surprisingly, Darwin - along with other evolutionary scientists - failed to consider what might happen if mutation affected the brain of our primate ancestor, as if humans were somehow immune from further change, after all, it's not just feathers or fat layers that can change.

31. The leverage affect prompted by a mutation to primate brain architecture expressed for 'recall' (called elsewhere in the academic literature 'episodic memory'), so humans were now able to provide *themselves* with the means for modifying nature, creating niches nature doesn't provide. Unique in nature, the ability to modify both the environment *and* their own behaviour, early

hominids were able to 'out-compete' not just close competitors, but *all* other species.

32. Recall supersedes the kind of memory that's 'prompted' by events from the environment – as thirst in an elephant increases its ability to detect the presence and the direction of a water source.

33. The elephant does not *think* " I'm thirsty", has no need to 'think' at all, but performs the

needs of being an elephant with nothing more than the carefully structured mental pathways of an elephant, clever as they are. And elephants are very clever at being elephants.

34.   Humans store and carry water. They plan to ensure a water supply, if necessary fashioning vessels to transport it. Humans can draw maps of where water is and another human will understand it. Standard theory fails to account

for this extraordinary evolutionary advance.

35. This quantum leap in mental diversity explains why humans now keep elephants in zoos, rather than the other way around. Thus, the concept of 'leverage' is at the heart of the Neoteric view of the proliferation of species.

36. Nor is there a progression from 'primitive' to 'more advanced' as we humans believe and as Darwin

implicitly accepted: bacteria are widely seen as primitive organisms. In fact they are the oldest surviving, having spent at least 450,000,000,000 (4.5 billion) years as an essential component of the biosphere. Without them we humans, and a host of other species, couldn't digest food. Bacteria 'drive' life on Earth.

37. Meanwhile, the supposedly most successful, *Homo*

sapiens, are arguably the most destructive.

38.   Recall is the key mental asset driving constructs of an abstract kind, independent of time or place. An example is the way a human – of any era - might mark a trail through the wild in order to find the route back; at intervals, grass or branches broken or bent, stones moved or upturned, to create a 'trail'. This structure has no relevance in the

environment, in Nature. The trail has no significance whatsoever to any other species . Only humans realise and appreciate its significance. Driven by recall, the relevance is 'sequence', 'order', 'abstraction', all uniquely present in humans and absent elsewhere in nature.

39.   Neoteric theory holds that this unique mental attribute separates humans from all others, not by virtue of a divine

instigator or by belonging to a superior kind of scientific priesthood. Instead, humans occupy a 'mental niche', using abstract concepts such as *time* and bring into being such notions as 'past', 'present' and 'future'. In this way, humans manipulate, via recall and anticipation, the outcome of events. Recall enabled us – uniquely - to produce the tools vital for this quantum leap in evolution; evolution *itself* evolved.

40. The complex process of anticipation, prediction and continued learning generates feedback from the mental niche which, in turn accelerates further development. In this way humans have emerged to 'out-pace', 'out-perform' and 'out-distance' all other species, to the point where all other species are now under threat from a 'mental niche'

expanding exponentially in nature.

41. A good example of the mental niche may be seen in large cities such as London; it continues to survive as an artificial entity despite outliving thousands of generations of humans living and labouring there.

42. An example of a 'pseudo species' may be further offshoots of human activity such as institutions or commercial groups that

'outlive' by many generations their original human founders. Pseudo organisms and the pseudo species exist in countless examples.

43. However it may not be the case that 'the world will just go on'. The biosphere is for the first time under severe threat directly from the first pseudo species; the human. While still physically a primitive ape, humans have come to dominate a small and insignificant

planet of rich - but finite - resources.

44. An inherent danger stems from the fact that there is apparently no upward limit on this exponential expansion and may, already, be beyond any constraints that might naturally occur.

45. The Darwin family suffered greatly; from the effects of intermarriage with close relations; loss and stress from reversal of faith;

anguish that came from the insights implicit in the Theory of Evolution. This was by no means the end of an expansion of countless negative outcomes that stem from the original mutation conferring recall. These were implicit in the first use of flint implements fashioned as weapons.

46. The current level of arrogance in scientific certainty, certainty in religious and political belief and the certainty that endless human

development is sustainable are all equally at fault in accelerating human ambition beyond the limits imposed by the physical resources available to us. Perhaps we have already trespassed beyond the capacity of the biosphere to repair itself; a point of 'no return'.

*NON FORAS IRE, IN INTERIORE
HOMINE HABITAT VERITAS*

The final argument...

4.5 billion years ago (?) the first signs of life appeared on Earth. Fossil evidence in ancient rocks of that date bears the traces of single cell life forms. They have been identified as early bacteria. (?) Evolution had already begun; life forms adapting and becoming more complex.

All life today has a precursor; a simpler, less complex form. A more complex form emerges, better adapted to changing conditions than the forms of an earlier epoch. So life develops in parallel; adaptations to the more complex exist in the present alongside their simpler predecessors. The chimp exists in the same world at the same time as human descendants – so does gut flora. Microbes, nearly identical to ancient life forms, dwell in the stomachs of contemporary apes and humans. Indeed, it is the tiny microscopic forms, hardly visible to the naked eye, that dominate the world. These are probably more representative of 'life' than any one of species so much more familiar to us; snakes, tigers, elephants, gorillas, chimps and buffalo.

The fact that we also share DNA with microbes, as we do with chimpanzees, is largely overlooked. But we are recent descendants of the primate family. Microbes are the most ancient of all life forms and they are still here, sharing the contents of our stomachs. This close association suggests another, more intriguing origin of speciation; perhaps bacteria themselves helped determine the path and the pattern of emerging genomes.

If this novel theory has any merit it is because it eliminates the need to introduce an external factor of causation, such as radiation emanating from the cosmos giving rise to mutation. Perhaps instead the subtle influence of bacteria, central as they are to their continuing existence, helped shape all life forms from the very beginning.

How this mechanism might operate is a mystery, but no more mysterious than the idea that any genome already stores genetic information that only needs to be triggered by mutation to induce a 'beneficial' change, such as bluer, more distinctive feathering.

What does the entire genome of any species actually contain? At the moment, most is a complete blank; perhaps 5% of coding can be directly linked to physical characteristics, very little to neurology, for example. Few species have been studied apart from humans.

The human genome project has achieved wonders; logging, recording data and describing chemical bonds representing the molecular components of the human DNA coding. Assigning function or meaning is a much harder task. Perhaps much of the as yet un-decoded data is historic record of previous incarnations; redundant now, but 'may come in handy' on the occasion of some future global upheaval.

In proportion to the size of, say, an orange, the atmosphere surrounding the Earth is comparable to the thickness of the tissue wrapping that sometimes accompanies that fruit; the atmosphere is about 8 miles thick, the limit of life on Earth.

Added to that, the oceans cover around seventy per cent of its surface. Much of the rest is mountain or desert. What is left has to support not only pasture and space for crops, forest and water reserves but also the growing extent of densely populated cities. The natural habitat of 10 million or more other *non-human* species is gravely threatened.

The chances of finding an alternative place to live

anywhere within reach of planet Earth are, to tell the bald truth, nil.

Evolution *itself* has evolved, and (in the case of human) is evolving at exponential, speed - faster than our biosphere can cope.

Darwin had not an inkling of this predicament.

Now we do.

www.ingramcontent.com/pod-product-compliance
Lightning Source LLC
Chambersburg PA
CBHW061231180526
45170CB00003B/1249